CORRALLING THE TROJAN HORSE

A Proposal for Improving U.S. Urban Operations Preparedness in the Period 2000–2025

Russell W. Glenn ▪ Randall Steeb ▪ John Matsumura

Prepared for the United States Army

RAND
ARROYO CENTER

The research described in this report was sponsored by the United States Army under Contract No. DASW01-96-C-0004.

ISBN 0-8330-2977-0

Published 2001 by RAND
1700 Main Street, P.O. Box 2138, Santa Monica, CA 90407-2138
1200 South Hayes Street, Arlington, VA 22202-5050
RAND URL: http://www.rand.org/
To order RAND documents or to obtain additional information, contact Distribution Services: Telephone: (310) 451-7002; Fax: (310) 451-6915; Internet: order@rand.org

PREFACE

The challenges of today's urban operations arise from the vast expanses of man-made structures, the tens of thousands of innocent civilians, and the need to preserve friendly force and noncombatant lives as well as physical infrastructure. Successfully addressing these challenges requires moving away from a reliance on the predominantly firepower-based doctrines of the past. This document provides one such alternative.

This research was sponsored by the Assistant Secretary of the Army for Acquisition, Logistics and Technology and was conducted in RAND Arroyo Center's Force Development and Technology Program. The Arroyo Center is a federally funded research and development center sponsored by the United States Army. This document will be of interest to individuals in government and commercial sectors whose responsibilities include doctrine, policy design, funding, planning, preparation, or the development of technologies in support of civil or military operations involving urban environments in both the immediate future and longer term.

For more information on the RAND Arroyo Center, contact the Director of Operations, (310) 393-0411, extension 6500, or visit the Arroyo Center's Web site at http://www.rand.org/organization/ard/

CONTENTS

SUMMARY

The United States stands at a difficult point in history, one in which implied strategic objectives and the nation's military capabilities are dangerously in tension. Policymakers in Washington, D.C. perceive a very low public tolerance for U.S. casualties during armed conflict. There is also evidence that Americans will no longer accept large numbers of noncombatant casualties during U.S. operations overseas. At the same time, there is an ever-increasing likelihood that the country's military forces will be committed to operations in urban areas. The world's population, economic wealth, and power are concentrated more than ever in cities, and this trend is only increasing. Combat in urban areas has historically been very costly in terms of casualties. But the problem is more complex yet. There is an inverse relationship between the amount of firepower used and friendly casualties: the more firepower applied to a built-up area, the lower the friendly force's losses. Unfortunately, those lower military casualty numbers come at the expense of higher numbers of noncombatants killed and wounded and vastly increased damage to civil infrastructure, a tradeoff that may or may not be acceptable depending on public tolerance and whether or not the city in question belongs to a friend or foe of the United States. The result is multiple conundrums: tolerance for friendly casualties is low while the probability of urban combat is high. The means of reducing friendly military casualties cause increased civilian losses.

This is, quite simply, an unsatisfactory state of affairs. While there has of late been renewed interest in military operations in built-up areas, progress is too slow. Better ways of conducting urban operations in the future are becoming increasingly feasible. The time for critical decisions on doctrine, component and system testing, training, and funding is at hand.

This documented briefing presents alternatives to continuing "business as usual." It is a product of ongoing work by the RAND Arroyo Urban Operations Team to find ways to accomplish missions in urban areas at reduced loss of friendly force and noncombatant life and with less damage to city infrastructures. This new approach is an alternative to today's "close combat" urban operations paradigm that requires commitment of large numbers of soldiers to urban actions. It has six mutually reinforcing, largely simultaneous components:

- Intelligence, surveillance, and reconnaissance (ISR) enhancements
- Denial of access
- Nodal operations
- Noncombatant control
- Selective dominance
- Post-conflict sustainment.

Though the discussion largely focuses on combat operations, these elements are also applicable to serving the objectives of stability and support missions, whether part of a larger combat contingency or the dominant actions in the theater of operations.

The period considered is 2000–2025. There is continuity inherent in the analysis. It is recognized that a goal of improving the conduct of urban operations demands somewhat consistent fiscal policies, technological development, and leader support over that span of years. After a brief consideration of what makes urban operations notably challenging, the authors review innovative concepts for enhanced urban readiness within the framework of the six components identified above. A brief conclusion considers obstacles to progress.

This nation's founding fathers gave the world a magnificent gift: a government and ethic that champions the rights and importance of man as an individual. Their example, the United States of America, greatly influences today's perceptions of what is right and wrong during the application of military force. That conflict should minimize the loss of friendly soldiers' lives and civilian suffering whenever possible is a noble and praiseworthy standard. It is also one that dramatically complicates military operations, for war is by nature a destructive undertaking, the more so when it is conducted in densely populated areas. Today the objectives of mission accomplishment, low military casualties, and minimal noncombatant loss are too often mutually exclusive. They can be less so in the future. Enhanced knowledge of friendly force, enemy, and innocents' locations and intentions would allow less random applications of military might. Separating combatants from nonfighters would largely remove the latter from harm's way. Focusing on only those physical and temporal targets of value to task achievement would allow other locations to go unharmed. Applying force with well-considered discretion, and doing so with an understanding that a successful transition to post-conflict life demands restraint in the application of violence, the U.S. armed forces can conduct successful operations in villages, towns, and cities. Doctrine requires modification. Training must improve. There is a need to guide and capitalize on technological developments. Organizations must adapt. This document provides one view on how America's military can attain these desirable ends.

ACKNOWLEDGMENTS

The authors thank A. Michael Andrews, Larry Stotts, and Pam Beatrice at ASA(ALT) for their continued support of RAND Arroyo Center research in the urban operations arena. ASA(ALT) was one of the first U.S. government agencies to recognize the increased importance of the subject area. Their support has been fundamental to the maintenance of ongoing initiatives to improve the nation's urban operations readiness.

The reviews of this document by LTC John Gordon (USA, ret.) and LtCol Duane Schattle (USMC, ret.) were notable for their valuable insights; they are most appreciated. The work done by Terri Perkins and Nikki Shacklett in preparing this document for publication was, as always, exceptional. Thanks to Priscilla B. Glenn for the outstanding cover design concept, and to Ron Miller, Benson Wong, and those others whose fine design and publication skills did so much to make publication possible.

GLOSSARY

C2	Command and Control
C3	Command, Control, and Communications
CEP	Circular Error Probable
C4ISR	Command, Control, Communications, Computers, Intelligence, Surveillance, and Reconnaissance
CINC	Commander in Chief
DOE	Department of Energy
EFOG-M	Enhanced Fiber Optic Guided Missile
FAE	Fuel Air Explosive
HUMINT	Human Intelligence
IPB	Intelligence Preparation of the Battlefield
ISR	Intelligence, Surveillance, and Reconnaissance
NGIC	National Ground Intelligence Center
NGO	Non-Governmental Organization
PGMM	Precision Guided Mortar Munition
PVO	Private Voluntary Organization
PSYOP	Psychological Operations
RF	Radio Frequency
ROE	Rules of Engagement
UAV	Unmanned Aerial Vehicle

1. Introduction and Background

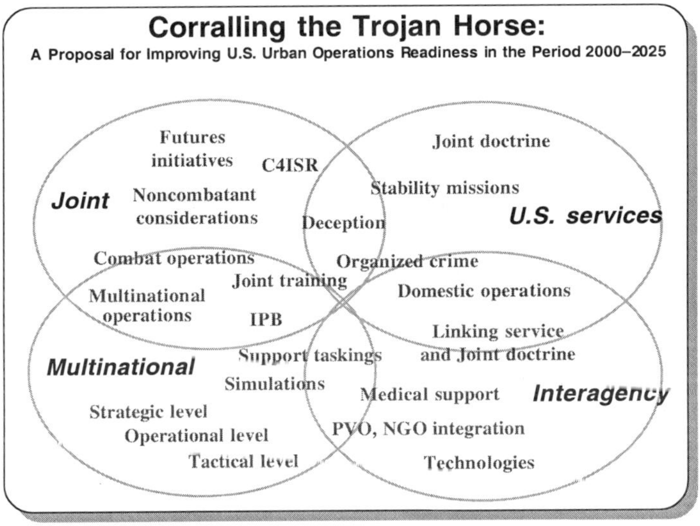

Military operations in urban areas are rarely "business as usual" for a military force or the government that oversees it. The variety of challenges includes confronting adversaries that deliberately endanger innocent civilians in hopes of reaping negative media coverage for U.S. forces (Beirut, 1982; Mogadishu, 1993), high-profile events such as the loss of aircraft to supposedly technologically inferior foes (Mogadishu, 1993), and the likelihood of casualties more numerous than normally suffered on other battlefields. The inherent physical complexity of built-up areas and their habitually high levels of human activity further complicate operations. In some cases the Trojan Horse is of the enemy's making; the cause of failure is a previously veiled capability or tactical advantage. Sometimes it is a force's own lack of preparedness or its failure to match military capabilities and strategic requirements that introduce the agents of disaster, as was the case when unexpected attacks on Hue and Saigon in early 1968 adversely influenced American public support for the Vietnam War, or when the loss of eighteen soldiers on October 3–4, 1993, precipitated a presidential decision to evacuate Somalia.

This documented briefing summarizes an initial effort by RAND Arroyo Center's Urban Operations Team to find ways to both enhance U.S. force readiness for urban contingencies and reduce the cost in lives of those actions. Research and analysis in this regard has been under way for over two years; it continues, both in support of better defining the proposal forwarded here and determining other feasible paradigms. What the reader will find in this document is a snapshot of a continuing conceptual evolution.

The current methods of conducting military operations in urban areas today differ little from those employed in 1968 Hue or 1993 Mogadishu. They in fact greatly resemble tactics and even operational-level approaches used during World War II. These methods have proved costly in human lives and physical infrastructure. Continued reliance on them is ill-advised given today's strategic environment, yet little is being done to significantly alter the ways the U.S. military conducts operations in cities.

This analysis considers alternatives to historical urban military procedures. Though the focus is largely on combat operations, there is considerable utility in applying the concepts we offer here to the full spectrum of potential future contingencies. Similarly, many of the proposals can be applied to areas other than cities. The various elements in the graphic on page 1 are a sampling of those that should be considered in any comprehensive viewing of urban operations. The diagram is demonstrative only, however, for were it a true Venn diagram virtually all the elements would have to fit within the intersection of the joint, U.S. services, multinational, and interagency regions.

The concepts that follow consider desirable changes to U.S. urban operations capabilities during the first quarter of the 21st century. Technological developments will be fundamental to the adoption of future approaches to urban operations, but progress will also demand changes to doctrine, training, and, possibly, organizations. Members of RAND Arroyo Center's Urban Operations Team have visited national laboratories, private commercial enterprises, and other entities in efforts to establish the feasibility of specific technologies and their developmental timelines. Thus our analysis includes no "and then a miracle occurred" leaps forward that do not have a basis in feasible technological development during the period of consideration.

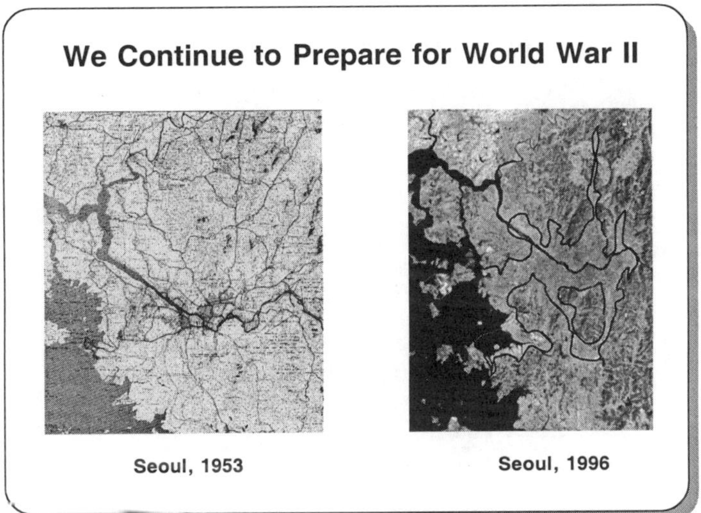

Photographic images of Seoul courtesy of the United States Army 33rd Engineer Detachment (Terrain), Yongson, Republic of Korea.

When North Korean forces attacked southward into the Republic of Korea in June 1950, the population of Seoul was one million persons, a figure that did not too greatly exceed the active duty strength of the United States Army at 691,000.[1] The city stood by itself, an urban island amongst acres of rice paddies and hills. Inchon was miles to the west, Suwon similarly distant to the south, and towns such as Uijongbu and Tongduchon stood well north. Today all of these are cities in their own right, and all are linked to each other by unbroken stretches of urbanization. (See area outlined in right-hand image.) Units the size of those that seized Seoul a half-century ago could easily disappear into any one of the many skyscraping apartment complexes that reach upward from the north bank of the Han River.

Despite these changes, history's lessons have much to offer. A commander who fails to isolate an urban area before attacking or who undertakes operations without synchronizing combined arms and joint assets does so at great risk. Yet in other ways the methods employed in conducting World War II and Korean War urban operations are in many ways no longer applicable; fire bombing and block-by-block annihilation of cities are now probably techniques only of the past for Western militaries. They will be even less tolerated two to three decades from now.

---

[1] Strength of U.S. Army from *American Military History*. Washington, D.C.: Center of Military History, 1989, 540.

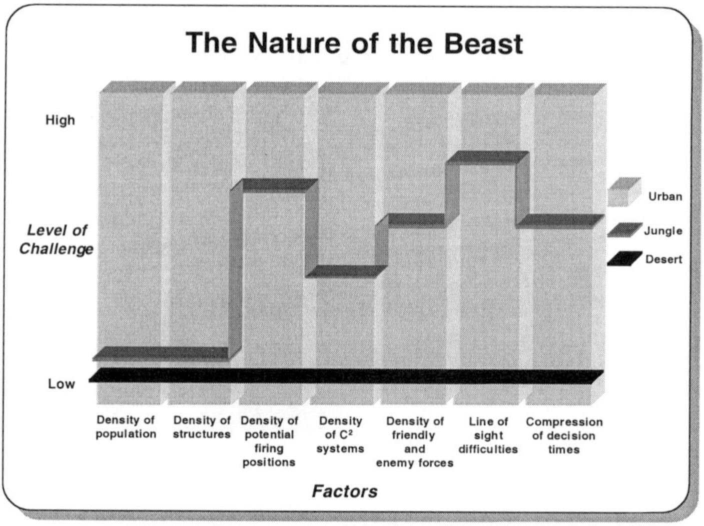

Solutions might be far easier to discern were urban operations difficult for just one or two reasons. But the challenges posed when a military force undertakes missions within a built-up area are varied, highly complex, and replicated nowhere else. Both the density of noncombatants and the sheer numbers of innocent civilians can overwhelm a force. The higher densities of many other relevant factors further complicate intelligence analysis and military operations in general.

Stringent rules of engagement and a desire to minimize loss of noncombatant life further complicate a unit's efforts to accomplish its assigned tasks. During urban combat operations, an adversary can choose from an extraordinary number of potential firing positions that can be at, below, or above ground level. Command and control are complicated by severe and repeated line-of-sight interruptions. Restricted line-of-sight can also result in many engagements at very short range; the average rifleman's engagement range in a city is well under 100 meters. Such close-in fighting tends to neutralize many of the technological advantages that Western nations' weapons systems offer in other environments.

Finally, these short ranges and the proximity of units to other friendly forces, enemy, and the seemingly ubiquitous noncombatants mean that decision times are often dramatically less than those in other types of operations. An operational-level reserve that would be tens or even hundreds of kilometers away on other terrain might be only blocks away in a large city. The possibility of its moving rapidly along through-building or subsurface approaches could severely reduce the time available for decisions at even the operational level.

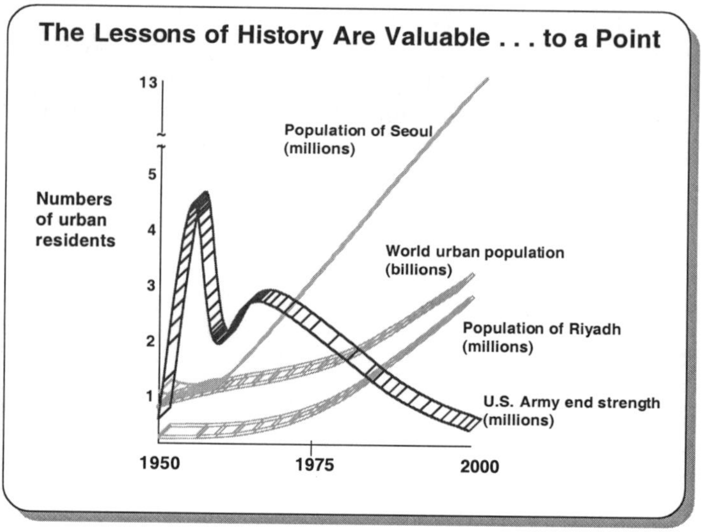

The scope of the challenge confronting today's militaries is suggested by the figure's depiction of changes in urbanization and force strength over the last half-century. The world's urban population has increased from less than a billion in 1950 to some three billion in the year 2000. Riyadh in 1950 was a city of 60,000; in fifty years its population has increased over forty-fold to 2.6 million.

The implications are multifold. An army having to fight in a modern megalopolis not only has the daunting task of confronting an enemy; the tasks associated with the control and support of noncombatants could easily demand more manpower than was employed to seize entire cities in the mid-20th century.

These massive increases in population have been matched by a similar inflation of the numbers of vehicles, streets, buildings, and rooms in cities, as well as the volume of three-dimensional space they encompass. (With the dramatic vertical growth of urban areas, it is more appropriate to speak of the volume encompassed than the amount of area covered when considering urban challenges.) The type of operation that characterized the World War II seizures of Manila, Aachen, and other urban entities is in many ways no longer feasible.

Military Victory = Strategic Loss?

Battle	Noncombatant KIA	U.S. KIA	Noncom-to-Friendly KIA Ratio
Manila (1945)	100,000 (est.)	1,010	100:1
Hue (1968)	5,800	150	11:1 (w/ 400 ARVN KIA)
Panama (1989)	202	26	8:1
Mogadishu (1993)	500+	18	30:1

- **The traditional definition of operational success may no longer apply. Instead:**

Success = Mission Accomplishment + Acceptable Friendly Casualties + Tolerable Noncombatant Casualties

Further investigation does little to mitigate concerns over the armed services' ability to successfully conduct urban combat operations. The four examples in the chart entailed commitment of American ground forces to close combat. In all instances, the result was significant friendly casualties. In the case of Mogadishu, 73 were wounded in addition to the 18 killed. Together these losses were a considerable portion of the total U.S. force directly involved in combat on October 3-4, 1993. (Note that the losses shown reflect only soldiers and civilians killed. Total casualties, including those wounded and missing, were considerably higher.) Notably, however, noncombatant casualties were considerably higher than friendly force losses. These examples, spanning nearly fifty years of recent history, lend credence to a conclusion that close combat may not be an effective alternative if success demands more than accomplishment of the "purely military" mission. Unfortunately, however, this is currently the only viable approach in virtually any near-term contingency.

The nature of close combat urban operations makes it attractive for adversaries seeking to neutralize an armed force's technological advantages. The United States has a dramatic asymmetric advantage over most potential adversaries on most types of terrain. Built-up areas are an exception. There, much of the fighting is infantryman versus infantryman, rifle against rifle, and entails sudden engagements conducted at very short range.

Though some enhancements in the next few years may slightly reduce the losses of friendly soldiers and marines, there is little promise that initiatives within the bounds of the close combat paradigm can significantly affect casualty levels. An alternative is to remove the soldier, marine, and noncombatant from the immediate environs of lethal urban combat to as great an extent as possible.

2. The Road Forward

Where Do We Go From Here?

- ***The Road Forward***
- **Steps on a Path Toward Strategic Success**
- **Potholes and Better Construction**

The remainder of this analysis considers the changes that are necessary to prepare U.S. armed forces for the inevitable urban operations to come. It presents a new approach, one that seeks to reduce the number of friendly forces that must be committed to the built-up area, and it suggests other changes to increase the probability of operational success. Finally, the briefing provides a list of obstacles that stand in the way of success.

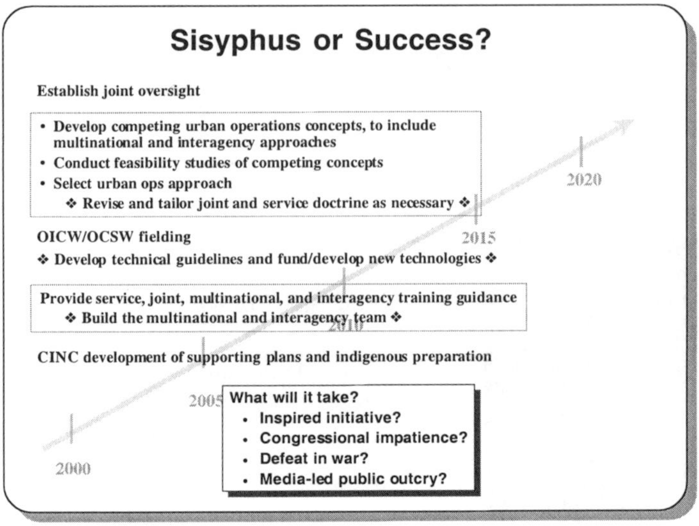

This chart only begins to address the actions that are flagstones along the path to operational preparedness. As has been noted, the challenges inherent in solving urban operations issues are very complex; finding solutions will be no less so.

It should be apparent that marginal changes to tactics and technological developments related to them provide too little improvement too slowly. Urban operations continue to pose the threat of very high casualties, making commitment to major undertakings in built-up areas a strategic conundrum: political or military objectives demand fighting in a city, but the high casualties that will follow could threaten to cause withdrawal of support for continued operations. Urban operations during Tet 1968 in Hue and Saigon and those in 1993 Mogadishu had strategic consequences of this sort. These and other historical examples suggest a need to pursue alternative approaches to conducting urban operations. Yet little headway has been made in this regard. Any alternative to the current method of committing large numbers of men to city interiors obviously demands not only discussion of promising technologies but also development of supporting doctrine, training, and organizations.

Ongoing improvements to doctrine represent but a beginning to what needs to be a continuous process. New doctrinal guidance will influence, and should in turn be influenced by, evolving technologies.

Throughout this process there is a need to seriously consider and, where appropriate, apply multinational and interagency doctrine to complement ongoing U.S. joint and service efforts. In addition, joint and service doctrines need to incorporate urban considerations throughout doctrinal manuals and training now and continuously in the years to come.

Other needs are also being addressed too slowly. Though the 1994 Defense Science Board called for a joint champion to oversee urban operations initiatives, none exists. Longstanding C4ISR problems that affect urban and other contingencies continue virtually unabated, many of them less technical than procedural issues or problems attributable to a lack of joint cooperation. Training facilities are too small for effective training at any but the lowest tactical levels. The lack of readiness to conduct urban operations is no secret, yet the topic still receives marginal attention in most services and joint commands. Barring a sustained armed services assault on this topic, it appears that pressures to change will have to come in the form of further military embarrassments or initiatives outside the armed forces.

3. Steps on a Path Toward Strategic Success

Where Do We Go From Here?

- **The Road Forward**
- ***Steps on a Path Toward Strategic Success***
- **Potholes and Better Construction**

A New Way of Doing Business, 2000–2025

Solutions will require a comprehensive approach:

- **Combined arms, joint, multinational, and interagency**
- **Include offensive, defensive, stability, and support components**
- **Operations may or may not include combat**
- **International and domestic operations**
- **Strategic, operational, and tactical levels of war**
- **Cover operations influenced by**
  - **Strict rules of engagement (ROE)**
  - **Concerns regarding casualties**
  - **Reduced force strengths**
- **Significant near-term improvements will be more influenced by doctrine and training than technology. In the longer term, technology may enable movement to a new urban operations paradigm.**

Any approach, whether evolutionary or revolutionary, must expand the scope of what is currently too often an overly narrow view of military actions in cities. Urban operations are not exclusively military, nor are they limited to combat actions, tactical activities, or international venues. The varied nature of past (and future) scenarios is one of many reasons that solutions to urban challenges are so extraordinarily difficult to develop.

Transition to a less costly urban combat paradigm will be evolutionary rather than revolutionary. Initial advances will be marginal and more the result of changes to doctrine and improvements in training than technological leaps forward. There are, however, a number of promising technologies that should, given steady development, funding, and a disciplined approach, provide the armies of 2010 to 2020 strikingly new ways to accomplish the missions inherent in urban contingencies and to do so with fewer friendly and noncombatant casualties.

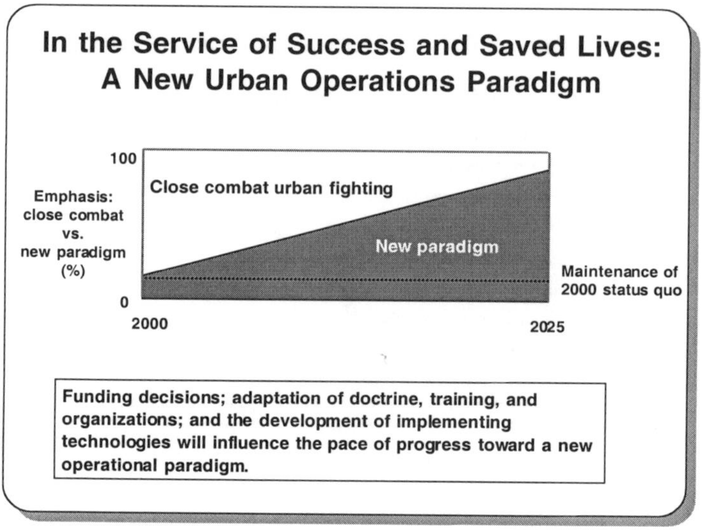

Finding an alternative to urban close combat is essential if Western armed services are to meet today's standards of operational success. While it is unlikely that a military will ever be able to completely avoid sending ground forces into a built-up area, their numbers (and thus the risk of large numbers of friendly force casualties) can be reduced.

Any shift will require:

- Commitment of resources
- Development and acquisition of new systems
- Testing, both with simulations and field analysis
- Training with adapted tactics
- Creation of fresh doctrine, and
- Systematic transitions to new force structures.

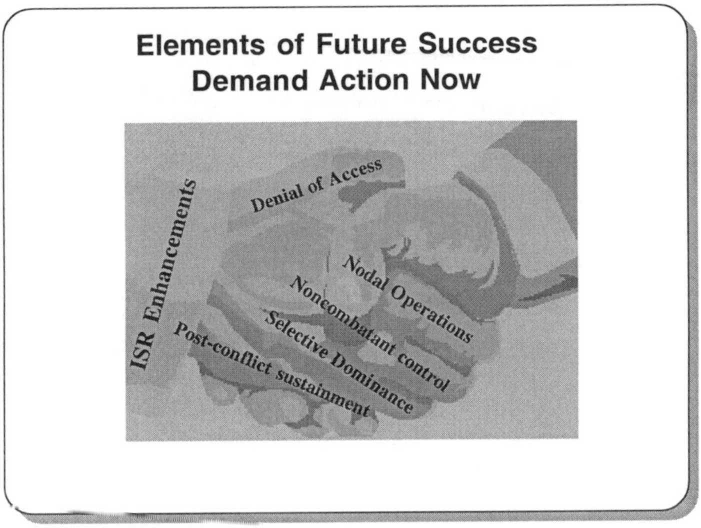

RAND analysis has determined that there may be alternatives to these historically costly close combat operations. The following several slides review what can be done in both the immediate future and in the longer term to move away from the short-range engagements that neutralize a more sophisticated force's technological advantages. The six components comprise an alternative approach. They should not be viewed as successive steps but rather as initiatives to be employed simultaneously. Their development should likewise be coordinated to maximize the synergy achievable through proper project management.

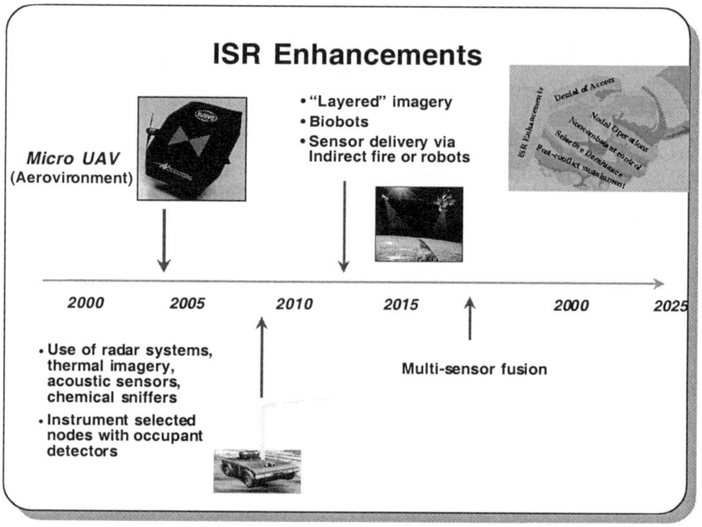

Images: Micro-UAV provided by Aerovironment Inc., Simi Valley, California. Robotic vehicle originally from *http://www.spawar.navy.mil/robots/land/mdars/mdars.html.* Earth imagery: *http://spaceflight.nasa.gov/gallery/images/shuttle/sts99/html/jsc2000e01556.html.*

The conduct of effective intelligence, surveillance, and reconnaissance (ISR) activities underlies successful urban operations today; it will continue to do so in coming years. Unfortunately, built-up areas greatly complicate these tasks. Masking or outright concealment provided by buildings and underground passageways currently shields adversaries from detection. The same problems hinder locating and identifying noncombatants. Underground facilities are difficult to detect; structure interiors are only partially discernable through the use of algorithms that consider external clues to internal layouts such as vent, door, and window locations; hazardous materials may be found in unexpected places.

Today, HUMINT plays a greater role during urban contingencies than is the norm in other operations. HUMINT collection poses special challenges, however, not the least of which is the time needed to establish a viable network. Technology offers some solutions in the long run. Very small unmanned aerial vehicles (micro-UAVs) should be able to fly within buildings or underground passageways and transmit images back to users. Combining visual, infrared, thermal, electromagnetic signature, nonreflective radar, other imagery or measurement capabilities, vibration measurement, and algorithms such as those mentioned above could one day allow three-dimensional and even interior mapping of specific urban areas of interest.

Developmental dates as shown on the timeline are estimates of when operational capabilities of such systems could be fielded given the current state of development. Fielding dates will of course be influenced by funding, the level of support given specific projects, and other variables.

While some technologies have shown less immediate promise than others, the same end may be achieved via different approaches. For example, knowing whether an enclosure (apartment, room, building) is occupied and whether or not any occupants are armed has great value. However, nonreflective radar of sufficient resolution and transportability may not be available for a considerable period. Possible alternatives that address the same or related requirements include:

- A means of temporarily sealing an enclosure without infantrymen having to complete the dangerous and both mentally and physically exhausting action of clearing each one.

- Chemical detectors, acoustic sensors, or another technology that provides another way of "seeing" through walls.

- It may at first sound absurd, but insects could be fitted with sensors or even small weapons, then guided to target areas through the use of pre-positioned or robot- or artillery-delivered bait.

Technologies offer considerable potential for the enhancement of urban ISR. Without a coherent and comprehensive approach, however, development is likely to be slow, poorly integrated, and late.

Denial of Access: Interdiction and Preparation of Indigenous Forces and Terrain

Prevent entry into urban area/Limit penetrations

- **Employ controllable vehicle barriers (barricades, superlubricants, foams, combustion inhibitors)**

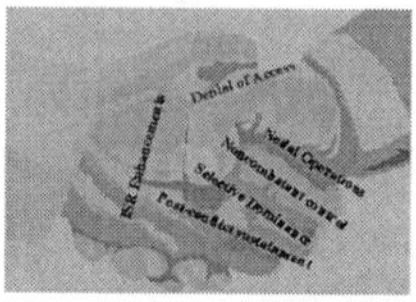

Train in-place forces

- **Prepare indigenous forces for tasks appropriate to their tactical, operational, and strategic missions**
- **Capitalize on niche capabilities and U.S. technological advantages**

Ensure infrastructure protection

- **Design protected systems to deny enemy control of water, power, gas, phones, and other critical facilities**
- **Use limited destruction versus complete annihilation**
- **Establish backup power and C3 redundancy**

Those structures critical to the safety and well-being of an urban area's residents should be protected if at all possible. There will also need to be backup generators, communication systems, emergency water supplies, and medical treatment facilities.

A key facet in protecting an urban area during conventional operations is the synchronization of obstacle networks, defending forces, and supporting air. Pop-up commercial vehicle barriers could be placed on avenues of approach to urban areas to complement others permanently in place and ready to be executed (such as those in South Korea). These commercial systems have the advantage of being mobile and thus could be deployed on city streets themselves.

The proximity of noncombatants to areas of likely contention reinforces the need for employment of nonlethal obstacles when feasible. Nonlethal obstacles such as superlubricants and quick-hardening foams can be located at key canalizing points to complement or serve in lieu of traditional obstacles such as ditches. Nonlethal weaponry has application beyond protection of noncombatant life, however. For example, commanders can employ selected information operations (to include jammers and RF bombs) and nonlethal systems as needed to disable enemy capabilities in proximity to significant infrastructure nodes or proscribed targets.

The training of indigenous forces to defend specific critical nodes or deny their use by the enemy could facilitate long-term mission success.

Subsequent arrival of reinforcements from the United States or coalitions would then focus on retention of such crucial assets, or, in cases where they were rendered inoperable, recapture and repair. Repair could be facilitated by clever initial planning. Removal of specific components that would be difficult for the adversary to acquire would allow returning friendly forces to quickly reinstall them and return the system to operation (given that the enemy did not otherwise damage the facility). Similarly, designs for facilities known to have strategic importance could include consideration of the need for rapid removal of components, much as bridges and other facilities in some nations were built with prechambers for destructive explosives. In other cases, limiting the extent of destruction or developing secondary assets could similarly assist in meeting strategic requirements.

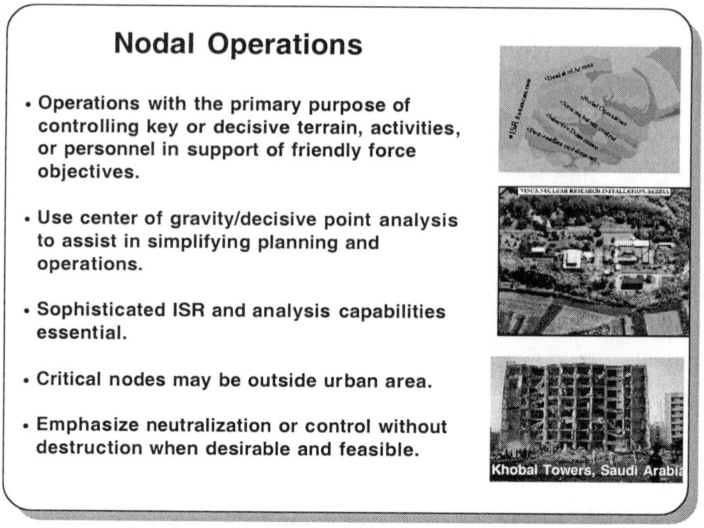

Image of nuclear research facility from *http://www.defenselink.mil/photos/May1999/990529-O-9999M-005.gif.*

Image of Khobal Towers damage from *http://www.defenselink.mil/cgi-bin/photoform.pl?photo=Jun1996/960626-N-00000-002.jpg.*

It may be difficult to determine which of the many military, political, infrastructure, and other of a city's nodes are vital to mission success. Using the Clausewitzean concept of center of gravity and that of decisive points[1] holds considerable promise in this regard. If wisely applied, these concepts ought to assist in making a daunting task more manageable. This is especially true in cases where a commander and his staff find themselves confronted by a modern megalopolis that encompasses thousands of cubic kilometers, tens of thousands of structures, and millions of people.

Including post-conflict population support and rebuilding requirements during planning further aids a commander in wisely selecting the means of "taking down" a target. A major difficulty in planning the destruction or termination of essential services is determining the second- and higher-order effects on friendly force missions and noncombatant survival. Sophisticated staff analyses, perhaps completed in conjunction with civilian support agencies, will be crucial. Such analyses must also include consideration of the

---

[1]Decisive points are points that have value due to their potential influence in unbalancing a center of gravity.

adversary's perspective, e.g., what are friendly-force centers of gravity and how can control of these and other vital key nodes be used against them? In addition, there may be essential services whose control is in enemy hands at the beginning of an operation. The British colony of Hong Kong, for example, received most of its water from the People's Republic of China. Any Cold War defense would have had to compensate for this limitation.

Noncombatant Control

Operations intended to influence noncombatant attitudes and behavior in a manner beneficial to friendly force objectives.

Current capabilities:

- **PSYOP**
- **Civil affairs**
- **Coercion (e.g., use force, amenities denial)**

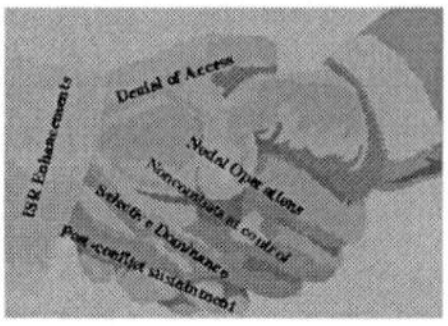

Emerging and future capabilities (a sampling):

- **Acoustic (noise), microwave, light, and combustion alteration technologies**
- **Foam, superlubricants, other barriers or surface treatments**
- **Embrittlement munitions**
- **Image and acoustic projections**
- **Calmatives**

Noncombatant control could take the form of ensuring the existence of marked routes to shelters, dispensing tags for noncombatants and friendly forces, or using calmatives or incapacitators for crowd control. The key is to influence behaviors in a manner conducive to friendly force objectives. Often this will include preserving noncombatant lives via precluding their presence in areas of high risk, e.g., locations of imminent friendly force attack.

Current capabilities require either willing cooperation on the part of civilians or the use of coercion. With the development of more effective nonlethal systems, both overt and covert means of rapidly influencing civilian behavior will be available.

Activities in support of such operations could include:

- Distribution of food, medical support, and other relief from locations outside of contested areas
- Use of psychological operations (PSYOP) and deception
- Use of nonlethal weapon systems to move noncombatants out of an area in which they could be used as human shields.

Selective Dominance

- **Controlling specific areas, activities, or segments of the population to facilitate overall mission accomplishment**
- **Demands innovative doctrine technologies:**
  - ***Controlled engagements***
    - **Highly accurate (small CEP)**
    - **Controllable effects**
  - ***Sector and seal***
    - **Limits adversary mobility**
    - **Denies observation/engagement of friendly forces**

Photographic image of Kosovo from *http://www.defenselink.mil:80/photos/Apr1999/990406-O-0000M-007.html.*

The previous four elements individually offer the potential for dramatic capability improvements. Together they provide a foundation block for a new urban operations paradigm, one in which combat operations are less dependent on commitment of friendly forces to city interiors. Through the synchronization of high-resolution intelligence, separation of noncombatants from the combat zone, isolation of the adversary, and the application of precision munitions, the commander will often be able to dominate selected parts of an urban area without physically occupying them. Allowed to concentrate instead on identifying and addressing those areas whose control is vital to mission accomplishment, he maximizes the benefits derived from employing highly accurate, remotely guided, controllable-effect munitions to minimize both friendly and noncombatant losses.

These controlled engagements would be all the more effective when used in conjunction with "sector and seal" capabilities designed to entrap an adversary or control his movement. Robotic delivery of foams to seal passageways, use of acoustic (noise) or microwave nonlethal systems, and remotely delivered lethal or nonlethal obstacles would act to fix, canalize, turn, or block forces that could subsequently be engaged. Nonlethal means would be used to segregate noncombatants from combat areas, further preventing inadvertent casualties and limiting interference with friendly operations.

Selective Dominance: Reducing the Need for Close Combat

Potential systems (lethal):
- Overpressure devices (FAE bombs)
- Laser-guided PGMM
- Urban EFOG-M
- Lethal robotic "pointman"
- Air platform-delivered small smart bombs

Nonlethal technologies are key:
- Jamming munitions
- Foam guns for sealing openings
- Acoustic (noise) weapons
- Microwave weapons

Image of Marines: *http://www.defenselink.mil/photos/Aug1996/960826-A-8072J-012.html.*
Image of MDARS-E courtesy of Space and Naval Warfare Center, San Diego, CA.

This chart shows a sampling of systems relevant to establishing selective dominance. The first employs an overpressure concept. Many countries have produced fuel-air explosive bombs that may be air or rocket delivered. They spread a mixture of fuel vapor over a large area and then ignite it, producing a strong shock wave. At close range, such munitions can disable weapons and kill soldiers. They can incapacitate at longer ranges or when effects are mitigated by obstacles. The effects may be reduced or magnified by structures, depending on the location of the target, building materials, and other factors. Fuel-air weapons come in a number of forms, from indirect-fire systems to the "Shmel," a shoulder-fired rocket launcher similar to an RPG in appearance.

Whether lethal or nonlethal, systems must possess the dual characteristics of accuracy and controllable effects. Systems launched from manned aircraft, ground, or sea platforms; laser-guided weapons; munitions flying preprogrammed flight profiles; or others deliverable by remotely controlled or preprogrammed robots could all meet these requirements. Cities put special demands on such capabilities, however. For example, an urban model of the EFOG-M would have to be able to turn sharply in narrow city streets or otherwise navigate so as to engage a target in a specific room without endangering nearby friendly forces or noncombatants.

As previously noted, nonlethal capabilities would help a force confronted with an enemy that deliberately exposes noncombatants to friendly fires or in cases where it is essential to influence behavior without employing lethal capabilities. It should not be forgotten that nonlethal means can also be valuable in enhancing the application of lethal force or in attacking other-than-human targets, e.g., the enemy's command and control systems.

Robotic ground platforms have special applicability to high risk/close combat tasks. Smaller and more stealthy than many manned systems, they should be more difficult to detect and destroy. They are in many cases less affected by nonlethal systems such as acoustic weapons and incapacitants. Unlike humans, they can be used to intentionally draw fire to expose enemy positions. However, there remain a number of policy issues regarding the incorporation of autonomous lethal capabilities in autonomous systems.

Image of Drmno thermal power plant, Serbia, from *http://www.defenselink.mil/cgi-bin/photoform.pl?photo=May1999/990527-O-9999M-004.jpg*.

Image of Smederevo highway bridge over Dunav River, Serbia, from *http://www.defenselink.mil/cgi-bin/photoform.pl?photo=Apr1999/990430-O-9999M-005.jpg*.

Soldier construction image from *http://www.defenselink.mil/cgibin/photoform.pl?photo=Oct1996/960920-A-9384H-012.jpg* .

Images of destroyed house and church under construction taken by the author in northern Bosnia-Herzegovina.

Continuing improvements in nations' abilities to control the effects of military operations means that objectives may increasingly be attainable without wanton destruction and extensive loss of civilian life. It is now often possible to neutralize an enemy asset without completely destroying the infrastructure on which it depends, as is the case when chaff is used to temporarily render a power distribution facility inoperable. Such methods may be complemented by other approaches in the future, such as specially designing facilities of notable strategic importance to allow the removal of key components, as was previously addressed. Care in planning and wisdom in future weapons development will speed post-hostilities recovery both because the original damage done was consciously limited and because support and rebuilding assets were positioned early. The result will be lives preserved, money saved, and a more rapid return to normalcy.

Image of Pyongyang, North Korea, from U.S. government files.

The implications of urban operations are far reaching. Future initiatives must include consideration of systems' capabilities to support urban contingencies. The reverse is also true: development of innovative tactical, operational, and strategic approaches to urban operations should incorporate consideration of other concepts. Light- and medium-weight vehicle development, for example, may considerably reduce U.S. military reaction times to overseas contingencies. Such improvements would significantly influence the nation's ability to assist allies; they would also impact on the character and scope of those allies' preparations for urban denial operations.

Better doctrine and improved training will provide the basis for most significant near-term improvements. Given proper recognition and funding, selected technologies will increasingly offer benefits for a military force seeking to improve its chances of success. As long as the United States relies on short-range close combat to achieve its urban objectives, however, a skilled enemy will be able to extract a steep price for such accomplishments. The solution lies in moving away from these close combat engagements if those objectives include dramatically lowering friendly and noncombatant casualties, reducing collateral damage to the infrastructure, and mission success.

4. Potholes and Better Construction

Where Do We Go From Here?

- **The Road Forward**
- **Steps on a Path Toward Strategic Success**
- ***Potholes and Better Construction***

Excuses

The problem is too complex

It cannot be solved on one man's watch...or two...or three

The warrior ethic *or*
"Close with and destroy the enemy" = manned aircraft = dead Americans

Issues are not in leader comfort zones

Let coalition forces do urban operations

Just don't do it — Avoid urban areas

Progress in improving urban operations capabilities has been too slow. The chart lists only a few of the many excuses for such lethargy. It goes without saying that preparedness for urban operations is not the only concern of the armed forces. However, recent history, world demographics, and the growing importance of cities and other built-up areas worldwide all point to the inevitability of further such actions in the future. History and common sense make it obvious that to continue to rely on close combat is a surefire way of ensuring high friendly and noncombatant casualties in the future. Tradition and old ways of doing business must be challenged in the interest of serving national objectives and saving lives. This briefing has outlined one; good analysis demands creation of others. If the United States is to move away from the close combat paradigm, there must be a presentation of competing approaches.